HEREDITY AND MEMORY

HEREDITY AND MEMORY

BY

JAMES WARD Sc.D.

PROFESSOR OF MENTAL PHILOSOPHY IN THE
UNIVERSITY OF CAMBRIDGE

BEING

THE HENRY SIDGWICK MEMORIAL LECTURE

DELIVERED AT

NEWNHAM COLLEGE, 9 *NOVEMBER* 1912

Cambridge :
at the University Press
1913

CAMBRIDGE
UNIVERSITY PRESS

University Printing House, Cambridge CB2 8BS, United Kingdom

Published in the United States of America by Cambridge University Press, New York

Cambridge University Press is part of the University of Cambridge.

It furthers the University's mission by disseminating knowledge in the pursuit of
education, learning and research at the highest international levels of excellence.

www.cambridge.org
Information on this title: www.cambridge.org/9781107425736

© Cambridge University Press 1913

First published 1913
First paperback edition 2014

A catalogue record for this publication is available from the British Library

ISBN 978-1-107-42573-6 Paperback

HEREDITY AND MEMORY

WRITING in the middle of the xvIIth century on "the efficient cause of the chicken," William Harvey of Caius College in this university quaintly remarked: "Although it be a known thing subscribed by all...that the egge is produced by the cock and henne, and the chicken out of the egge, yet neither the schools of physicians nor Aristotle's discerning brain have disclosed the manner how the seed doth mint and coin the chicken out of the egge." "How much nearer a disclosure," asks a writer who quotes this passage, "are we to-day?...On the whole we have [still] to confess that we do not know the secret of development, which is part of the larger secret of life itself[1]."

[1] J. Arthur Thomson, *Heredity*, 1908, p. 416.

Now it is commonly taken for granted that on this great problem—the problem of Heredity—psychology can have nothing to say. But I have come at length to think that, provided we look at the world from what I would call a spiritualistic and not from the usual naturalistic standpoint, psychology may shew us that the secret of heredity is to be found in the facts of memory.

But first of all, in accordance with the observation just quoted, it seems desirable to enounce a few general propositions true "of the larger secret of life itself" and applicable also to this particular part of it.

To begin—we find the processes observable in the world around us can all be ranged in one or other of two classes, as either anabolic processes or katabolic—to use in a somewhat wider sense the terms of a Cambridge physiologist. The former we take to imply action contrary to, the latter action along, the line of least resistance. The processes of the one order build up: those of the other level down. The one order implies that direction, in the sense of aim or end, which we associate with mind as sensitive and purposive; the other that indifference which we associate with matter as lifeless and inert.

Of such guidance or direction—I would ask you next to note—we have immediate experience only in the case of our own activity, as in building a house or organising a business. It may well seem rash therefore to attribute such processes as the formation of chlorophyll in a blade of grass or of albumen in a grain of corn to guidance in this sense. But at all events they are processes pertaining exclusively to living organisms, and found nowhere else. If these processes

should some day be artificially repeated in
a laboratory, as Professor Schäfer so confidently expects—even this would imply the
guidance of the living chemist. But still, it
may be asked, what right have we to identify
life and mind; what right, for example, to
credit plants with souls, as Aristotle did?
The right that the principle of continuity
gives us. No sharp line can be drawn between plants and animals nor between higher
animals and lower.

But here the advocate of Naturalism may
intervene. "Continuity is just as complete
regarded from below as it is regarded from
above," he may urge; "and if so, surely the
proper method of investigation is to begin
with the simpler and earlier rather than with
the later and more complex." Not necessarily,
we must reply : all depends—as Plato pointed
out long ago—upon where the characters we

have to study are clearest and most distinct. In the case of life there can be no doubt that this is where they are the most developed. "It is clear," as G. H. Lewes once said, " that we should never rightly understand vital phenomena were we to begin our study of life by contemplating its simplest manifestations...we can only understand the Amoeba and the Polype by a light reflected from the study of Man[1]." Moreover if we begin from the material side we must keep to this side all through : if matter is to explain life at all, it must explain all life. But it is evidently impossible to describe the behaviour of the higher organisms in physical terms. Indeed the ablest physicists recognise that the concepts of physics are inadequate to the description of life even in its lowest forms. We may conclude then, that when, as in the

---

[1] *Problems of Life and Mind*, 3rd Series, 1879, p. 122.

case of life, we are seeking to interpret the
meaning of a continuous series we must start
where that meaning is clearest, where it is
best known and most definite, not where it
is least known and most inchoate.

Working in this fashion from ourselves
downward, what then, we may now ask, do we
find as the irreducible *minimum* which all
life implies; and what are the most general
characteristics that mark every advance from
lower forms to higher? To the first question
the answer, I trust, may here suffice that life
everywhere implies an individual and an
environment. To changes in this environment the individual's behaviour is, or tends
to be, so adjusted as to secure its well-being.
For—it is worth remarking by the way—
throughout the realm of life the category of
value, which such terms as well-being and
ill-being imply, is relevant, though irrelevant

everywhere else. It is this that gives to what we have called guidance or direction its motive and its meaning.

As to the second question—what are the general characteristics of advancing life?—this is on the whole admirably dealt with in Herbert Spencer's Psychology under the heading *General Synthesis.* Defining life as the adjustment of internal relations to external relations, he shews how, as evolution progresses, this adjustment extends in range both spatially and temporally; how at the same time it increases in speciality and complexity; how separate adjustments are more and more co-ordinated and unified; and so forth. The interval between the palaeolithic man, whose world was bounded by his river-valley, who knew little of the past and planned less for the future, who lived from hand to mouth content with raw food and possibly no clothing, who possessed only the rudest implements and had but a cave for his dwelling—the interval, I say, between him and civilised man with his long traditions, fixed laws, innumerable arts and organised division of labour, sufficiently illustrates the character of this progress when already far advanced. The inheritance of the permanent achievements of one generation by the next is obviously the main factor of such social progress: this we may call heredity in the literal sense. But we talk also of heredity as a factor in the biological progress from the *Protista* up to Man; though here the heir and the inheritance can only be distinguished by calling the individual the heir and his organism the inheritance, that is to say by regarding as two what the biologist conceives as one. It is this biological heredity that is our problem.

Before attempting to attack this problem directly there is however still a characteristic of life or experience that we may for a moment consider, which seems to throw some light upon it. Experience I once proposed to define as the process of becoming expert by experiment; and recent elaborate observations[1] of the behaviour of the *Protozoa* shew that these microscopic creatures frequently succeed, as we do, only by way of trial and error. Even plants prove capable to a great extent of accommodating themselves to changes in their environment. All this presupposes a certain plasticity, which in turn implies retentiveness: in other words, just as later generations inherit from earlier generations so later phases of the individual inherit, as it were, from earlier phases. In

[1] H. S. Jennings, *Contributions to the Study of the Behaviour of the Lower Organisms*, Washington, 1904.

our own case we get a good deal of insight into this process in what we can observe of the growth of habit. What was originally acquired by a long series of trials and failures, engrossing all our attention, becomes at length "secondarily automatic"—to use Hartley's now classic phrase. Of this such feats as skating or piano-playing are familiar examples. This mechanization of habit is aptly described in the saying that "use is second nature." It sets attention free for new advances which would else be impossible. So *natura naturata* is the condition of further *natura naturans*. The organism gives us a warrant for the term "mechanization" in the permanent modification of brain and muscle which the acquisition of new dexterities entails; and *mutatis mutandis*, the same holds true of the knowledge we know so thoroughly that, as Samuel Butler said, we have ceased to know that we know it at all. Now this law of habit we may reasonably regard as exemplified in the life of every individual in the long line of genealogical ascent that connects us with our humblest ancestors, in so far as every permanent advance in the scale of life implies a basis of habit embodied in a structure which has been perfected by practice.

And now, starting from the analogy just noted—namely that habit connects successive phases in the life of one individual as heredity connects successive stages in the development of one race—we may pass at length to our problem. That analogy suggests the possibility of an indefinite advance upwards in the scale of life without the succession of individuals which heredity involves—provided, of course, that a single individual lived sufficiently long and did not grow old.

In place then of the innumerable individuals a certain genealogical ascent has successively entailed, let us imagine one individual accomplishing the whole of it. The final result as regards structure might conceivably be substantially the same, nor need the time required be very different. But there would certainly be one very important difference. For the solitary immortal without ancestry structure would be wholly the result of function. But for the many mortals—who have a racial history as well as a personal history—function would be the result of structure, so far, that is, as the embryonic stage of their existence is concerned. To this stage Haeckel accordingly gave the name of *palingenesis*; because in it the structural acquisitions of our imaginary immortal are recapitulated. But whereas the same level of development might have been attained in approximately the same time in the two cases—that of a single persisting individual and that of a continuous succession of perishing individuals—the recapitulatory process, peculiar to the second case, proves to be vastly more rapid. It took thousands of years, say, to produce the first chicken, but the hen's egg reaches the same level in three weeks. To be sure the recapitulation is not precise and complete in every detail; yet most biologists, I believe, allow that it is very full and substantial. Nevertheless even in the most complex organisms far the greater part of it is accomplished within a year. How may we account for this extraordinary brevity in the repetition of a process that took so long at first ? The consideration of this question in the light of our preliminary investigation may lead us towards a solution of our problem.

Let us suppose our imaginary individual, when he had proceeded but a little way in the slow and arduous fashion of a pioneer, to have been set back to the beginning once more, *still however retaining the memory of his former experiences.* "We may be sure," as I have said elsewhere, "that in that case he would make good the ground lost in much less time than he required at first, and also without following all the windings of the tentative route into which his previous in-experience had led him : his route the second time would be routine[1]." So, for example, a man, who had gradually by various improvements adapted a house or a machine to his purposes, would proceed, if either by some accident was destroyed. Let us again suppose that after a while, when a still further advance had been achieved, our imaginary

---

[1] *The Realm of Ends*, 2nd edn, 1912, p. 209.

individual had to submit to a like setback once more; and that after a yet further advance the same thing recurred again, and so on indefinitely. The result would clearly be that the latest acquisitions would be the least straightened out, the least automatic and the least fixed. Now we know directly by observation that in like manner the memories and dexterities that are acquired latest are the least engrained and the most likely to fail. And turning from the individual to the race, Darwin has shewn that here it is specific characters, which are acquired later, rather than the generic, upon which they are superposed, that are peculiarly liable to variation.

Now the situation we have supposed closely resembles that of a new complex organism. Though not identical with the old it is nevertheless continuous with it, and it does reproduce the ancestral acquisitions with the less hesitation or variation, the more they have become habitual, secondarily automatic or organised. This no doubt implies— *prima facie,* at least—that acquired characters are inherited. But it implies also—*a point too often forgotten*—that we should not expect any clear manifestation of such heredity till the functions that have led to structures have passed far beyond the initial stage where conscious control is essential to their performance.

This doctrine of the inheritance of acquired characters, enounced, I believe, by Aristotle, accepted alike by Lamarck and Darwin, was only seriously called in question some thirty years ago. But so rapidly has opinion swung round, that now the great majority of biologists—and especially of zoologists—reject this hypothesis, as they call it, altogether. If they are right, it is useless attempting to develope further the psychological theory I have been trying to suggest. At this point then we must pause to examine their objections.

First, it is said, neither observation nor experiment has so far yielded any really decisive evidence for the old theory of inheritance. But it is equally true that they furnish no conclusive evidence against it. There are at present no crucial instances either for or against it; but as I have just said, we should not expect that there would be. Even the theory of natural selection was not established in that way: there too the argument depends entirely on cumulative evidence and general considerations. It must be owned that a vast mass of worthless cases of hereditary transmission has been

exploded by Weismann and his followers;
and this has naturally led to a general
distrust of the rest. Yet, as a singularly
fair-minded and acute biologist, Delage, maintains, the evidence is still formidable; for as
another Lamarckian has said, "transforma-
tion...acts *as if* the direct action of the
environment and the habits of the animal
[the parent animal that is to say] were
the efficient cause of the change, and any
explanation which excludes the direct action
of such agencies is confronted by the difficulty of an immense number of the most
striking coincidences[1]."

Yes, the Weismannians reply, we allow
that appearances all point in the Lamarckian direction, but inasmuch as the *modus
operandi* of the transmission is altogether

[1] Professor W. B. Scott, *American Journal of Morphology*,
1891, p. 395.

inconceivable, we decline to believe that they are more than appearances. This is their second argument. But as the "impassioned controversy," as Delage calls it, has gone on, this argument, it is hardly unfair to say, has changed its form. No process of transmission being conceivable, it is assumed that no such process is possible. In this way the Neo-Darwinians relieve themselves of the difficulty there is on their side of proving a negative. But assumption is not argument, and our mere ignorance of the "how" alone will not justify us in rejecting *prima facie* evidence. We do not brush aside the facts of gravitation because we are utterly ignorant of the process which they involve.

We certainly are entirely in the dark as to how structural changes in the body of the parent can affect at all the structure of the germ which it nourishes; since the two are anatomically entirely distinct. But it is something to the point to note, as Cope did years ago, that there is at least one case of a very precise connexion between two distinct tissues which is perhaps quite as wonderful as the connexion between body-plasm and germ-plasm and hardly less mysterious—viz. the adjustment of skin-colouration to ground-surface brought about through the organs of sight. Of this the chameleon furnishes the most familiar but not the most impressive instance. I came the other day across an account of some experiments that seem clearly to imply the intervention in some way of consciousness, in bringing about this adjustment—an intervention which Cope surmised but could not prove. Into a tank of flat fish, whose colour matched its sandy bottom, a number of pebbles of a different colour

were introduced. As seen by the fish the
mosaic so produced would appear more or
less foreshortened; but presently for all
that the fish became mottled like the bottom,
not as it appeared to them at rest, but as it
would appear to an observer looking down
from above, like the enemies the fish had
to elude. Also it is perhaps noteworthy that
there is at least a very intimate connexion
one way between body-plasm and germ-
plasm, as the extirpation of the reproduc-
tive glands shews.

But in the third place not only is the
inheritance of acquired characters said to
be without specific evidence, not only is it
declared inconceivable and so incredible
a priori, but in what he calls "germinal
continuity" Weismann claims to have found
direct and positive proof that it is actually
impossible. At the same time he contends

that the mingling of ancestral characters in sexual reproduction is sufficient to provide endless variations on which natural selection can work, thus rendering it also needless. Continuity of germ-plasm in some sense is an obvious fact in any case; and Harvey's position, *omne vivum e vivo*, is one that no biologist is concerned practically to deny. What Weismann means by germinal continuity, however, is such an absolute continuity of the germ-plasm as entails its absolute discontinuity from the body-plasm. Let the changes acquired by the parents be what they may, they can, he maintains, make no difference to the offspring. The circular character of the argument is here again apparent. Till the impossibility of the Lamarckian position is proved, germinal continuity or continuity of the germ-plasm may exist, but the absolute isolation of this

germ-plasm from the body-plasm is still open
to question; and apart from this isolation
the Lamarckian position remains tenable.
In fact Romanes, who has criticised Weismann at length with much acumen, represents
him not as attempting to prove but as simply
"*postulating*" the absolute non-inheritance
of acquired characters and deducing the
absolute continuity of the germ-plasm in
his sense from this.

A brief examination of Weismann's main
position and a word or two on an important
supplementation, to which he was eventually
driven, may I trust suffice to bring us round
again to the theory of "organic memories"
or "engrams[1]" to which our preliminary

[1] This theory was first definitely broached by Professor Ewald
Hering in a lecture, *Concerning Memory as a general function
of Organized Matter*, delivered at Vienna in 1870.   It has
recently been revived and developed by Professor R. Semon of
Munich (to whom the word 'engram' is due), in *Die Mneme als
erhaltendes Prinzip im Wechsel des organischen Geschehens*,
2te Auf. 1908.

propositions led up. In order to see more clearly the issue raised, we may begin by comparing our imaginary immortal with some mortal double, who has reached the same level of biological development within a few short years. The bodies of both are related to the simplest form of life—the one directly the other indirectly—the one having developed from such a form by continuous interaction with its environment, the other, according to Weismann, having developed from a continuous stock of germ-plasm, entirely cut off from the environment since the unicellular stage. Now in both cases there is an "immortal" concerned—our imaginary individual in the one case, and Weismann's hypothetical germ-plasm in the other. Also the result attained is in both cases the same; and certainly it is the most wonderful instance we know of what we call an end, a τέλος. The fundamental factors in

the one case are (1) the environment, and
(2) experience: the latter an essentially
positive or teleological factor, since it is
always directed towards self-conservation or
betterment. But the factors fundamental in
the other case are (1) natural selection, and
(2) amphimixis or the periodical blending of
two ancestral strands of the immortal germ-
substance; and both alike are non-teleo-
logical.

Can it be that one and the same end can
be reached by such disparate means? Its attainment through experience without natural
selection is conceivable: its attainment without experience, by natural selection and for-
tuitous variation alone, is surely inconceivable.
That both experience and natural selection
have co-operated seems indeed to be the fact,
as Darwin himself at all events assumed. But
this co-operation Weismann with Wallace

and the rest of the Neo-Darwinians denies.
And this denial it is that entitles us to say
that *dis*continuity, or isolation from the body-
plasm, is the essential feature of Weismann's
germ-plasm. This germ-plasm persists indeed
continuously; but it persists, so to say, under-
ground, screened like the sacred queen-bee[1]
from all the functioning with the external
world which the offsets it periodically throws
out discharge. Nevertheless it alone—not
intercourse with the environment—is supposed to determine their structure. Can the
manifold adaptations such structures display
and the endlessly diversified series of them—
continuously advancing in complexity of
adaptation—which the evolution of life as a
whole displays, can all these, we ask again,
be explained from such a standpoint?

[1] An illustration used by Dr Francis Darwin in his Presidential
Address to the Brit. Assoc. 1908.

Yes, they can, was Weismann's answer: which, however, as I said, he afterwards qualified. Any detailed examination of his theory would require a volume, has in fact occupied many volumes. But the discontinuity just referred to—duly considered—suffices, I think, summarily and yet successfully to dispose of it. Let me try to make this clearer. First of all I must ask you to note that, so long as we confine our attention to the unicellular organisms, there is no special problem as to the inheritance of acquired characters, nor in fact as to inheritance at all; for, as Weismann himself has shewn, there is here no natural death, and therefore—strictly speaking—no succession of generations. In his own words, "Natural death occurs only among multicellular organisms, the single-celled forms escape it." Referring to the well-known cell-division by which the one individual becomes two, he remarks: "In the division the two portions are equal, neither is the older nor the younger. Thus there arises an unending series of individuals, *each as old as the species itself, each with the power of living on indefinitely, ever dividing but never dying*[1]."

And now, in the next place, "it should also be borne in mind that," as one of Weismann's ablest supporters has observed, "many of these unicellular organisms...are highly differentiated, i.e. [are provided] with great complexity of structure...and that many have very definite and interesting modes of behaviour, such as swimming in a spiral, seeking light or avoiding it...trying one kind of behaviour after another—functional peculiarities, some of which cannot be described without using psychological

[1] Italics mine.

terms[1]." Lowly as these forms of life appear
when compared with our own, still the
enormous diversity to be found among them,
their wide range in space and their prodigious
antiquity, together suggest that between the
highest of them and the very beginnings of
life lies a long, long history of "one kind of
behaviour after another." In the course of
that history functions were mechanized and
structures fixed.

All this while the situation is that of our
imaginary immortal developing by intercourse
with its environment, save that here the immortal is real and not imaginary; and also,
instead of remaining solitary, is continually
manifolding itself into new individuals who
start equipped with all its acquisitions. All
this while too the whole organism of every
individual is in touch with the environment

[1] J. Arthur Thompson, *Heredity*, p. 33.

3

and every new acquisition is passed on to all the individuals into which any one is manifolded. If we like to call this inheritance, then it is inheritance of acquired characters, and there is no other. And all this while too there is no ground for distinguishing between body-plasm and germ-plasm; unless the cell nucleus—without which the cell never divides and which always divides with it—were to be called germ-plasm on this account.

But now so soon as we advance to the *Metazoa* and the *Metaphyta*—that is to the multicellular animals and plants—we are supposed, according to Weismann, to be suddenly confronted with an absolute discontinuity of mortal body and immortal germ. The one can no longer bequeath, the other can no longer inherit. Surely in such a case what we should naturally expect would be not the enormous advance from jelly-fish to mammal, which is what we find, but rather a practically stationary state. For consider the parallel case of social progress referred to earlier. The main factor in such progress we found was heredity. Imagine then, if you can, what would happen if now from this time forth every new generation had to begin where the old *began* and not where it left off; if no single human product from now onwards outlasted the individual who produced it; if in short all tradition and inheritance were from henceforth no more. Such a breach of social continuity between the future and the past would surely be startling; and yet it is an even greater discontinuity that Weismann seemed to imagine as marking off the *Protozoa* from the *Metazoa*. The Lamarckian factor good up to that point has, in his view, been absolutely inoperative beyond it. Natural selection and the blending of ancestral traits, more or less diverse, have henceforth sufficed.

The essentials of this process of amphigony or mixed generation, as Weismann conceived it, are simple enough. Two packs of shuffled cards he has himself suggested as representing the paternal and maternal germ-plasm. Now imagine half of each rejected and a new pack made of the remainders. A portion of this new pack will remain unaltered, though it will grow indefinitely: this portion is the latent germ-plasm that continues undeveloped until the advent of a new generation, when the processes of mixing, shuffling, and reducing division are repeated. From the remaining portion the new individual arises. Any one of the cards would suffice for its complete development, for each corresponds to a single ancestor. But there are many of them, and so a struggle ensues. The card or plasm of one ancestor—a paternal grandmother say—succeeds in shaping the nose: another —perhaps a maternal great-grandfather— provides the mouth, and so on. Though any one of them might conceivably have furnished the whole, it is far more likely that all or most of the pack have a share in it. In all this—the mingling of ancestral plasms, the reducing division, the reversion and atavism—we are, so far, in the region of fact. But when Weismann asks us to regard these ancestral plasms as themselves quite stable, we enter the region of fable. And even in this region, if we try to picture out—not everyday instances of heredity— but the primitive heredity, the link, that is to say, which long ago connected the

evolution of mortal multicellular organisms
with the wholly distinct evolution of immortal
unicellular organisms, we are utterly at a
loss to find any resemblance between the two.

No doubt amphimixis is a potent source
of congenital variations, and such variations
are the indispensable *point d'appui* for natural
selection. But what avails that if all possible
variations are confined to such unit-cha-
racters as the unicellular organisms display ?
As Henry VIII long ago remarked: "You
can't make a silk purse out of a sow's ear,"
turn it how you will. And how, in view of
the stability of germ-plasm and its discontinuity from body-plasm—which, according
to Weismann, are maintained throughout
multicellular evolution—the higher levels of
life were reached is a mystery. For, as Delage has said :—"Without the inheritance of
acquired characters there can be no new

ancestral plasms, and without ancestral plasms more complicated than those of the *Protozoa* there can be none of the superior animals[1]." But Weismann rejects the inheritance of acquired characters and so cuts off the first possibility: also he maintains— or rather began by maintaining—the absolute continuity and isolation of the germ-plasm, and so cut off the second.

Simultaneously with Delage's criticism however—in 1895 that was—Weismann made the complete change of front to which I have just alluded, abandoning—or at least essentially modifying—both his main positions, viz. the stability of germ-plasm and

[1] Yves Delage, *L'Hérédité et les grands Problèmes de la Biologie générale*, 1903, p. 560. But to Professor Hartog of Cork belongs the credit of first conceiving "cette objection capitale sous la forme du dilemme," as Delage expressly allows. *Op. cit.* p. 559 n. See Professor Hartog's letter in *Nature*, vol. xliv, 1891, entitled, "A Difficulty in Weismannism," pp. 613 f., and a second letter, vol. xlv, pp. 102 f.

its discontinuity from body-plasm. In the same year, H. F. Osborn[1] had said : "If acquired variations are transmitted there must be some unknown principle in heredity, if they are not transmitted there must be some unknown factor in evolution." "A perfectly correct conclusion" Weismann at once replied, and set to work to find the new factor by the simple method of extending the range of natural selection to the ultimate constituents of the germ-plasm itself. He supposed that the drama of the world without is here repeated on a minute scale. The determinants or ultimate constituents of the germ-plasm struggle with each other for nutriment. Some at length succumb and the successful survivors are thus "selected." So variations arise within the germ itself,

[1] In a lecture entitled *The Hereditary Mechanism and the Search for the Unknown Factors of Evolution.*

followed of course by corresponding variations in the body that they eventually build
up. This, his latest theory, the theory of
intra-germinal selection, has practically con-
vinced nobody; though it has been hailed
by opponents of Darwinism as sounding the
knell of the theory of natural selection[1].
Why, even if we allow the propriety of
regarding Weismann's hypothetical determinants as organisms struggling for nutriment—and facts seem to be against it[2]—

---

[1] "With a 'rehabilitation' of natural selection in the real
Darwinian meaning and only fair application of the phrase the
new theory has nothing whatever to do. It is, much more,
a distinct admission of the inadequacy of natural selection to do
what has long been claimed for it." Kellogg, *Darwinism to-day*,
1907, p. 199.

[2] "Actual experimentation on the influence of food-supply in
development does not bear out the assumption on which the
theory of germinal selection rests. Weismann himself gave the
larvae of flies...an abnormally small food supply..., with the only
result that the mature individuals were dwarfed; that all their
parts were reduced in size, but the actual proportions...were

why, even then, it *must* follow that useful and
consentient variations will appear just when
and where they are wanted, is more than
Weismann has seriously attempted to prove;
though he confidently asserts that it is so.
What is important about his new theory
however is the surrender both of the ancestral continuity and also of the somatic
discontinuity of the germ-plasm, a surrender
that, as Delage and many others have urged,
undermines his whole position. In short,
while the ground on which was based his
direct and positive proof of the impossibility
of the inheritance of acquired characters, is
abandoned, his full and definite admission
of the need for some equivalent of that
Lamarckian factor remains. We may then
now resume our consideration of what I have

unchanged." Kellogg, *Op. cit.* p. 201. But cf. especially Plate,
Die Bedeutung des Darwinischen Selectionsprincipes, 2te Auf.
1903, pp. 164–70.

called the psychological or mnemic theory
of heredity. Obviously we must begin from
the biological side, for we have no *direct*
experience in the matter. Moreover it is
not *our* memory, but a so-called organic
memory with which we are concerned.

Now if it is once admitted that the body
influences the germ-plasm in one way—i.e.
by way of nutrition—the possibility of its
influence in other ways can hardly be denied.
Even the influence in the one way may mean
a great deal more than Weismann surmised,
may involve not merely quantitative differences but qualitative differences as well.
Truly, as Mephistopheles said : "*Blut ist ein
ganz besondrer Saft.*" For example, through
the blood there circulate certain secretions,
called hormones, destined—as our Regius
Professor of Medicine has said—"for the

fulfilment of physiological equilibrium."
"Thus," as he goes on to say, "the reci-
procity of the various organs, maintained
throughout the divisions of physiological
labour, is not merely a mechanical stability,
it is also a mutual equilibration in functions
incessantly at work on chemical levels, and
on those levels of still higher complexity
which seem to rise as far beyond chemistry
as chemistry beyond physics[1].

But now the most striking instance of the
equilibration of its functions that an organism displays is that which brings about the
adjustment of internal relations to external
relations. In this adjustment indeed we may
say, as I have already urged, that life essentially consists. But this adjustment in the
higher organisms, where the characteristics

[1] Sir T. Clifford Allbutt, *Ency. Brit.*, vol. xviii, art. Medicine,
pp. 57 f.

of life are most distinct, is directed and sustained by a nervous system. Now to this system belongs in a pre-eminent degree that retentiveness and modifiability which life and experience everywhere imply—characters which we figuratively express by reference to plasticity, as in such terms as protoplasm, idioplasm, germ-plasm and the like. Plasticity, in a word, though most pronounced in the higher organisms, where the physiological division of labour has developed a distinct nervous system, is present in all organisms: all alike, though in divers degrees, live and learn. But now a multicellular organism, it is generally allowed, originated in, and still consists of, a colony of unicellular organisms —a colony that has become more of a commonwealth the further its functional division of labour has advanced. It is important then to insist on two points. First, every living cell, whether living in isolation or as a member of a complex organism, must be credited with that "organic memory" which all life implies. Next, we can set no limit to the consentient interaction between such cells when they have become "members one of another." "Even in normal circumstances" —to quote Sir Clifford Allbutt once more— "their play and counterplay, attractive and repellent, must be manifold almost beyond conception."

Now the reactions of the body-plasm in the higher organisms are, as said, guided mainly by the nervous system; and there, we know, facility and familiarity become automatic, more or less "unconscious" through repetition. But in the germ-plasm the rôle of regulating ontogeny, it is allowed, belongs mainly, if not solely, to the cell-nuclei or chromatin, Weismann's "germinal substance." "May we not therefore consider it probable that the nucleus plays in the cell the part of a central nervous system ?" was the question raised by Dr Francis Darwin in his masterly Presidential Address to the British Association in 1908. I am aware of no smallest detail in which the analogy between the two fails, and—in my opinion at all events—Dr Darwin was amply justified in contending, as Hering in his classic paper had done before him, "that ontogeny—the building up of the embryo—is actually and literally a habit."

Further, as Dr Darwin goes on to remark, this so-called mnemic theory is "strongest precisely where Weismann's views are weakest—namely in ·giving a coherent theory of the rhythm of development." The prodigious complication of Weismann's germ-plasm with its idants ids, determinants and biophores—

a vast army without a general—only spells confusion, when we try to picture it in motion, a plight like that of the builders of Babel after the confusion of tongues. According to the mnemic theory, on the other hand, the germ-cell is a definite unity, the counterpart of the structural alterations wrought by habit in the parental organisms with which it has been in sympathetic *rapport* all along. It is potentially what Leibniz would have called an *automate spirituel ou formel*, the latent entelechy of a future organism. We may compare it to a company of actors awaiting behind the scenes the call to begin their play. Each one knows his part by heart and also knows his cue. The routine or orderly rhythm of the performance is thus ensured and the play, though continually condensed at one end and extended at the other, has been so often repeated as to be acted without hitch or hesitation save perhaps in the cases of its latest amendments.

And now two or three last remarks by way of summary and conclusion.

No doubt when we try to ascertain the details of this process the difficulties in our way are, as Dr Francis Darwin candidly allows, "of a terrifying magnitude." But in all exploration the first thing is to secure, if possible, a general survey, a bird's eye view of the whole. If we are to see the wood, we cannot be among the trees. Now it is entirely with this preliminary problem that we have been concerned; and as befits such an inquiry we have tried—not to scrutinize the details of life—but to look at it as a real, concrete whole. Such a whole implies continuity: absolute breaks are impossible. Leibniz's maxim, *Natura non facit saltus,* is ours too. We find then no ground for separating organic life from psychical life: for us all life is experience. We cannot therefore assume that experience has no part in the building up of the organism, and only begins when a viable organism is already there. For us, ontogeny and heredity are aspects of a single process—a process that only experience will explain.

Again, the principle of continuity forbids us to assume that this process, by which an organism is built up, abruptly changes when we pass from unicellular organisms to multicellular. The way of trial and error and eventual success—function determining structure—followed in the earlier stages of the progress of life—in phylogeny, as it is technically termed—has equally been the way of its progress ever since. Every organism has proceeded from an organism ; yet among

unicellular organisms though there has been
progress there has been no genealogy. Among
multicellular organisms we find both : while
the offspring is still unicellular at the start,
the parents from which it sprang are unicellular no more. The greater this difference
in complexity between the offspring and its
parents, the longer the way its common ancestors will have traversed phylogenetically
or historically, and the longer too the way
that it will have to traverse ontogenetically
or automatically, before attaining to the
parental level and beginning a personal
history of its own.

This pre-natal, so to say, prehistoric life,
is called a heritage. And why? Because,
looking broadly at the whole record of multicellular genealogies, there appears to be
everywhere more or less correspondence and
nowhere a positive deviation between the two

itineraries—let us say—the historical and the automatic, the route of the original struggle and the routine of its recapitulation after many repetitions. But the more repetitions the more fixity. In short, what habit is for individual life that is heredity for racial life.

But the one lags enormously behind the other: the repetitions that will suffice to make a habit automatic for a lifetime are very far from sufficing to ensure heredity for future generations. And the more advanced the race, the farther heredity lags behind acquisition. At the unicellular stage, they are on a par, in other words there is as yet neither distinction nor interval between body and germ. At the multicellular stage there is both, and so, as the scale of life rises farther, the greater becomes the disparity between the still unicellular germ and the mature organism of ever increasing complexity; the greater therefore is the number of intermediaries through whose hands—so to say—the acquisitions of the one have to pass before they can be imparted to the other. And not only will these engrams —as they are called—be fainter on this account, and so require more repetitions to give them any permanence; but also for the same reason they will lose in definiteness and detail. Language, for example, will be inherited not as speech but as mere tongueiness and babble; art not as technical skill but as mere handiness or pure mischief; and so on. But if there is evidence of such inheritance of forms of behaviour impossible of acquisition at the unicellular stage, can we say that the continuity between body-plasm and germ-plasm ceased then? Can we at all understand such facts without recognising this continuity still?

But what exactly is this continuity between body and germ, and how are new acquisitions passed on? The continuity is what it always was, the continuity of membership in a commonwealth, where the whole is for the parts and the parts for the whole[1]; where all are more or less *en rapport.* The key to all this is to be found, I believe, in social intercourse, not in physical transmission. Unhappily however—as it seems to me—most of those who uphold the mnemic theory of heredity seem to hanker unduly after a physical explanation of the *modus operandi.* Hering and Haeckel talked of peculiar vibrations: if they were writing to-day they would probably refer to wireless telegraphy, as Professor Dendy has

[1] A Protozoan, we must remember, though unicellular, is far from simple, and an absolutely simple form of life is beyond our ken.

recently done[1]. But it is meaningless to talk
of memory unless we are prepared to refer
it to a subject that remembers. Records or
memoranda alone are not memory, for they
presuppose it. *They* may consist of physical
traces; but memory, even when called "unconscious," suggests mind; for, as we have
seen, the automatic character implied by
this term "unconscious" presupposes fore-
gone experience. But it is possible that a
subject may impart his knowledge or dex-
terity to another without dragging his pupil
through all the maze of blundering which
his own acquisition cost. The mnemic theory
then, if it is to be worth anything, seems to me
clearly to require not merely physical records

[1] *Outlines of Evolutional Psychology*, 1912. Signor E.
Rignano, founder of the international review *Scientia*, in particular
has developed a very elaborate theory of "organic memory" to
explain the inheritance of acquired characters, *Sur la transmis-
sibilité des caractères acquis*, Paris, 1906.

or "engrams" but living experience or tradition. The mnemic theory will work for those who can accept a monadistic or pampsychist interpretation of the beings that make up the world, who believe with Spinoza and Leibniz that "all individual things are animated albeit in divers degrees." But quite apart from difficulties of detail, I do not see how in principle it will work otherwise.

Milton Keynes UK
Ingram Content Group UK Ltd.
UKHW032321161024
449665UK00001B/16